# DIVING
## into Darkness

# DIVING
# into Darkness

## A SUBMERSIBLE EXPLORES THE SEA

## Rebecca L. Johnson

Lerner Publications Company ▪ Minneapolis

The author would like to extend special thanks to the following people: Dr. Craig Young, who invited me to join the May 1988 research cruise and explore the deep in *Johnson-Sea-Link II;* Dr. Lane Cameron, who introduced me to many bottom-dwelling animals and helped extensively with the deep-sea photographs; Phil Santos, who patiently answered my many questions about the *J-S-L* submersible; the members of the science party, who generously shared their knowledge of the ocean's inhabitants; the submersible pilots and crew; Captain Daniel Schwartz and the crew of the R/V *Seward Johnson;* and the people at HBOI, especially Pamela Blades-Eckelbarger, Tom Smoyer, and Susan van Hoek.

Library of Congress Cataloging-in-Publication Data

Johnson, Rebecca L.
    Diving into darkness : a submersible explores the sea / Rebecca L. Johnson
       p.    cm.
    Includes index.
    Summary: Discusses how scientists use submersibles to explore the depths of the ocean, surveys the development, operation, and equipment of the Johnson-Sea-Link sub, and describes a typical dive to the bottom of the sea.
    ISBN 0-8225-1587-3 (lib. bdg.)
    1. Oceanographic submersibles—Juvenile literature.
[1. Submarine boats. 2. Underwater exploration.] I. Title.
GC67.J64   1989                    88-27154
551.46'0072—dc19              CIP
                                    AC

Manufactured in the United States of America

1  2  3  4  5  6  7  8  9  10  99  98  97  96  95  94  93  92  91  90  89

# CONTENTS

# *INTRODUCTION*

If you have ever walked along a seashore or even looked at pictures of the ocean, you may have wondered what lies beneath the surface that seems broken only by endless waves. What kinds of creatures swim or drift just out of sight? How deep is the water, and what lives on the distant bottom?

Many other people have felt this same fascination with the sea. In fact, human beings have been asking questions about the oceans of the world for thousands of years. Ancient peoples sailed the seas and used them as a source of food and other useful products. But their ideas of what lay beneath the water's surface were surrounded by mystery and superstition.

What little knowledge these people had of the marine environment came from plants and animals that washed up on shore or were brought up in fishermen's nets. No one actually knew how these organisms lived or interacted with each other. By diving, humans could venture a short distance beneath the waves and briefly observe sea creatures in their natural surroundings. But even the best divers could hold their breath for only a few minutes before the need for air drove them back to the surface.

The main reason why the world's oceans remained a mystery for such a long time is that, until recently, there was simply no way to explore them. The ocean is an alien environment in which humans are slow and awkward invaders. Not only must we constantly breathe air, but without some kind of protection, we cannot tolerate the great pressures that increase dramatically as we go deeper.

**Modern scuba gear has made it possible for humans to venture a short distance beneath the surface of the sea.**

It is only in the last hundred years or so that we have begun to discover the true nature of the oceans that cover nearly three-fourths of the earth. We now know that beneath the ocean's surface lies a complex undersea world of soaring mountains and vast plains. This marine world is populated by an almost incredible variety of living things.

How have people learned about the ocean depths, and how will our knowledge of the undersea environment be increased in the future? Special equipment has made it possible for us to explore the ocean and to observe its inhabitants firsthand. This equipment includes devices that allow us to breathe underwater, to remain submerged for long periods, to withstand crushing pressures, and to descend to the sea bottom and come up again. In this book, you will learn about one such device, a deep-diving **submersible**. A submersible is a small underwater vehicle that can safely carry people far beneath the ocean's surface to study and explore the marine environment.

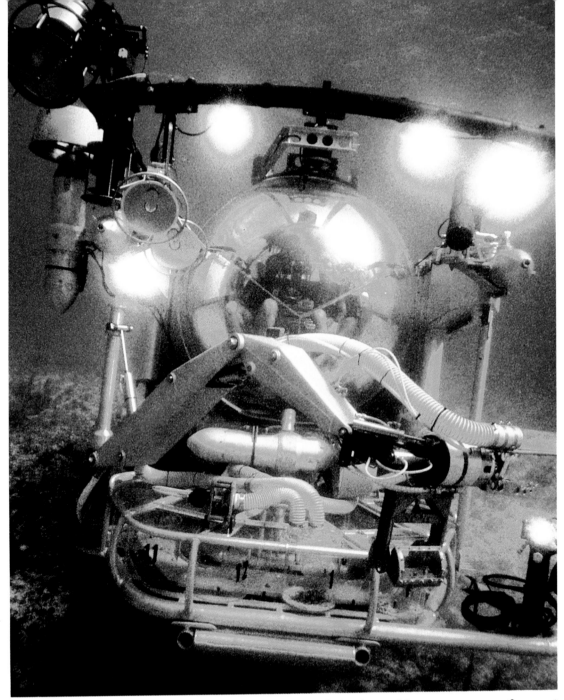

**Research submersibles such as the *Johnson-Sea-Link*, pictured here, are among the most versatile tools that scientists use today to explore the marine environment.**

According to legend, Alexander the Great was one of the first humans to explore the undersea world with the help of a special device. This Indian painting from the 1500s shows Alexander being lowered beneath the waves inside a glass "barrel."

# 1

# *EARLY ATTEMPTS TO EXPLORE THE SEA*

Research submersibles are a fairly recent development in the history of ocean exploration. In centuries past, many creative people wrestled with the challenge of designing underwater devices that would make it possible for humans to explore the deep. According to legend, the conqueror Alexander the Great may have been one of the first. Around the year 333 B.C., he was supposedly lowered beneath the waves of the Aegean Sea in a "barrel" made of glass. He returned from this underwater adventure with stories of strange creatures and fantastic sights.

In 1531, the inventor Guglielmo de Lorena designed a "diving bell" that was used to explore sunken Roman ships in one of Italy's large lakes. The diving bell was a small device that fit snugly around the diver's upper body. In front, it had a glass-covered opening through which the diver could see. But de Lorena's diving bell was so small that it must have contained a very limited air supply, allowing the diver only a brief stay underwater.

As the years passed, other inventors made many improvements on the design of diving bells. Air hoses attached to a bell, for example, brought fresh air down from the surface and made it possible for divers to remain underwater for hours at a time. Housed inside these contraptions, however, divers could travel underwater only as far as their connections to the surface would allow.

An illustration from an English magazine of 1752 showing how diving bells could be used to salvage goods from a sunken ship. On the right, a large bell shelters two divers. Fresh air, brought from the surface in a barrel, is piped into the bell through a tube. On the left, a diver standing on the sea floor wears a small bell over his head. He receives air through a tube connected to the larger bell.

For this reason, other people concentrated their efforts on designing underwater devices that would allow people to move more freely through the seas, unhindered by chains, air hoses, and metal helmets. In 1620, a Dutch inventor named Cornelius van Drebbel built one of the world's first workable submarine boats, which he tested in England's Thames River. Van Drebbel's device consisted of a wooden frame enclosed in layers of greased leather. This early submarine was propelled by oars that protruded from the boat's sides. It even had an air purification system that made it possible for the occupants to remain submerged for several hours.

Inventions such as diving bells and leather-covered submarine boats may seem strange and even a bit comical today. But such devices were important steps in the early attempts to explore the oceans. Not all these attempts were successful—there were often disappointing, and sometimes fatal, failures. Despite the dangers, many people, from scientists and inventors to explorers and even treasure hunters, continued to search for ways in which to explore the marine environment.

## A HALF MILE DOWN, AND BEYOND

In this century, there has been more progress made than ever before in developing the means to enter and investigate the oceans. The accomplishments of a number of 20th-century underwater pioneers have changed the course of ocean exploration dramatically. For example, up until 1930, scientists and explorers had penetrated only the first several hundred feet of the ocean. In that year, however, all previous depth records were broken by William Beebe, a biologist, and his colleague Otis Barton, an engineer. These two men were lowered to a depth of 1,428 feet (435.3 meters) off the coast of Bermuda inside a spherical underwater observation chamber. The thick-walled steel sphere, called a **bathysphere**, was designed and constructed by Barton.

In 1932, Beebe and Barton broke their own record by descending to 2,200 feet (670.6 meters) in the bathysphere. They did it again two years later, this time journeying to 3,028 feet (922.9 meters) beneath the surface—more than a half mile down! During these record-breaking

Otis Barton *(left)* and William Beebe pose with their two-ton steel bathysphere before making a record-breaking dive in August 1934. Beebe's vivid accounts of their experiences in the deep sea inspired later generations of ocean explorers.

dives, Beebe and Barton proved that the deep sea was not devoid of life, as many people had thought. On each dive they saw a great variety of animals swim past the bathysphere's two small windows. Some were familiar, but others were completely unknown.

Less than 10 years after Beebe and Barton's half-mile descent, another breakthrough occurred in ocean exploration.

In 1943, the undersea explorer Jacques Cousteau (then a young French naval officer), together with Emile Gagnan, an engineer from Paris, perfected a device they called the Aqua-Lung. The Aqua-Lung was the first practical form of scuba (self-contained underwater breathing apparatus) to be widely used. Divers wearing an Aqua-Lung carried their own air supply with them underwater. They could move about freely and could remain submerged for as long as their air supply lasted. Divers had never before experienced such freedom to swim through the sea. They were still limited in how deep they could dive, however, because the effects of pressure became dangerous beyond a certain depth.

Two other deep-sea pioneers achieved a long-awaited goal in 1960 by traveling to the bottom of the Mariana Trench, the deepest known place in the ocean. Jacques Piccard, a Swiss ocean explorer, and Lieutenant Don Walsh, a young U.S. Navy submarine officer, accomplished this feat in the *Trieste*, a deep-diving submersible known as a **bathyscaph**. After descending straight down for more than five hours, the *Trieste* settled into the soft muddy bottom of the Trench at the remarkable depth of 35,800 feet (10,911.8 meters). At this depth, nearly 7 miles down, the water outside was exerting 7 tons of pressure per square inch (about 1 metric ton per square centimeter) on the submersible.

The *Trieste*'s history-making dive marked the end of a 10-year "race" involving several countries and many types of vessels to see who could be the first to reach the ocean's deepest point. With that challenge behind them, ocean researchers turned to other goals. They began to spend their time and energy designing underwater devices that could greatly extend our ability to live and work in the marine environment.

One underwater pioneer of the 20th century who took this approach to deep-sea research was Edwin A. Link (1904-1981). Ed Link's name may be less familiar than those of William Beebe or Jacques Cousteau. But Link's many contributions to marine science and ocean engineering helped to bring undersea exploration to the point where it is today. In the following pages, we will take a brief look at these contributions and at one of the most important chapters in the modern history of deep-sea exploration.

**Ed Link's ship *Sea Diver* arrives in Jamaica to explore the sunken city of Port Royal in 1959.**

# 2

# EDWIN A. LINK: UNDERWATER RESEARCH PIONEER

Ed Link was undoubtedly one of the most ingenious people to become involved in deep-sea research in this century. Link was both an inventor and an engineer. In 1929, at the age of only 25, he made a name for himself by designing a flight simulator. This was a device on which new airplane pilots learned to fly before actually going up in planes. "Link trainers," as they were called, were used by the military to train thousands of aviators before and during World War II.

After the war, Link became very interested in underwater archaeology and turned his attention from the skies to the seas. Aboard his ship *Sea Diver*, he sailed the world in search of unique underwater finds. For nearly a decade, Link, his wife, Marion, and a team of divers explored the sunken remains of everything from early sailing ships to lost cities. *Sea Diver*'s deck was crammed with diving equipment and special machines for bringing objects to the surface. Among the artifacts that Link and his team recovered were 2,000-year-old ceramic jars, ships' anchors, coins and medallions, old cannons, and carved blocks of stone and marble. Many of these objects were of great historical importance.

17

## THE DECOMPRESSION PROBLEM

As the years passed, however, Link became more and more frustrated by the fact that many of the wrecks he wanted to explore were in water deeper than 100 feet (30.5 meters). This presented a problem: whenever divers spent more than a few minutes in deep water, they needed to undergo a period of **decompression** to counteract problems caused by exposure to the pressures at great depths.

In shallow water, a diver wearing scuba gear can remain submerged for more than an hour without experiencing any ill effects when he or she returns to the surface. As a diver descends into deeper water, however, the pressure that the water exerts on the diver's body increases dramatically. For example, at a depth of 67 feet (20.4 meters), the pressure is three times what it is at the surface. At 100 feet (30.5 meters), it is four times the surface pressure, and at 200 feet (61 meters), it is seven times surface pressure!

Such great pressures can have significant effects on a diver's body. For instance, if a diver returns to the surface

Holding onto a rope, a scuba diver makes a decompression stop while ascending to the surface after a deep dive.

18

too quickly after a deep-water dive, the rapid decrease in pressure can cause tiny bubbles of gas to form in blood vessels, tissues, or joints. This condition, known as the "bends," is extremely painful and can cause crippling or even death.

In order to avoid getting the bends after a deep dive, a diver must "decompress" by returning to the surface very slowly, making frequent stops along the way. Whenever Link's divers worked in deep water, they were forced to spend many minutes, even hours, slowly ascending through the water as their bodies adjusted to the pressure changes on the way up. Not only were these periods of decompression tiring and time-consuming, but the long, slow ascents also made the divers more vulnerable to attack by sharks.

# THE SUBMERSIBLE DECOMPRESSION CHAMBER

Ed Link was convinced that there must be a safer and more efficient way for divers to work in and explore the ocean depths. What was really needed, he decided, was a means by which divers could remain underwater for long periods of time. Imagine how much could be learned about the sea and its inhabitants if people could stay submerged for days or weeks! It was at this point that a plan began to take shape in Link's mind for a set of underwater "tools" that would make it possible to explore the oceans as never before.

As the first step in his plan, Link set about designing a submersible chamber that could carry divers down and up through the water, rather like an underwater elevator. After the chamber was lowered to a given depth, its interior pressure could be adjusted to correspond to the pressure of the water outside. Once the pressures were equalized, a diver wearing scuba gear could leave the chamber to complete underwater tasks or explore the area. The diver could return to the chamber to rest after each such excursion into the sea. When the underwater mission was finally complete, the diver would seal the chamber and ascend to the surface. Later, aboard *Sea Diver*, he or she would undergo gradual decompression in the safety and relative comfort of the chamber.

Link's **submersible decompression chamber**, or **SDC** for short, was ready for testing in 1962. Measuring 10.5 feet (3.2 meters) long and weighing 4,200 pounds (1,905.1 kilograms), the shiny cylindrical capsule floated upright in the water, with an entrance hatch on the bottom end. Hoses and power cables connected the SDC with its mother ship, *Sea Diver*. It could be raised or lowered along a length of chain that ran from the ship's deck down to an anchor on the bottom.

Link's submersible decompression chamber was the world's first undersea station. Like all new devices, it needed to be carefully tested. The first tests were carried out in relatively shallow water, and Ed Link was his own "guinea pig." On one occasion, he remained inside the chamber for 14 hours. For the first 8 hours, the SDC was submerged in 60 feet (18.3 meters) of seawater. After the chamber was hauled back aboard *Sea Diver*, Link spent another 6 hours inside, undergoing decompression.

The SDC performed well in its shallow-water tests. The time had come to go much deeper. In September 1962, history was made when Robert Stenuit, a young Belgian diver, descended in the chamber to a depth of 200 feet (61 meters) off France's Mediterranean coast, where he remained for more than a day. During his underwater stay, Stenuit left the chamber several times to explore his surroundings and perform various tasks. After each trip out into the water, he would return to the chamber, a cramped but welcome refuge beneath the waves.

Finally, after spending 26 hours below, Stenuit was brought back up to the surface and the SDC was secured on *Sea Diver*'s deck. Stenuit was in fine shape after his underwater adventure. But he was not overjoyed at the prospect of undergoing decompression inside the SDC for the next 65½ hours!

## A HOUSE AT THE BOTTOM OF THE SEA

Inspired by the success of his submersible decompression chamber, Ed Link moved on to the next stage of his plan for the exploration of the deep sea. Now the SDC would descend much deeper, and this time it would be accompanied by Link's newest invention, a black-and-yellow underwater "house" called **SPID (Submersible Portable Inflatable Dwelling)**. SPID was constructed of a weighted iron

**Ed Link's submersible decompression chamber is lifted into the sea from the deck of _Sea Diver_ at the beginning of Robert Stenuit's historic dive in September 1962.**

frame to which was attached a sausage-shaped, inflatable rubber balloon large enough to house at least two people. The entire structure rested firmly on the ocean floor. Divers entered and exited the inflated balloon through a hatch on the underside.

In July 1964, off Great Stirrup Cay in the Bahamas, SPID was lowered 432 feet (131.7 meters) to the sea bottom. Robert Stenuit and marine biologist Jon Lindbergh (son of Charles Lindbergh, the famous aviator) followed in the submersible decompression chamber. When they reached the bottom, the two men left the SDC, swam a short distance to SPID, and clambered up inside.

For the next 48 hours, the small

inflatable house was their home at the bottom of the sea. They ate, slept, and relaxed in its relatively warm, dry comfort. Periodically they went outside to perform various tasks and to investigate the dimly lit depths around them. What a strange feeling it must have been to live so far down, with a panorama of underwater life literally at your doorstep!

After spending two days underwater, Stenuit and Lindbergh returned to the submersible decompression chamber for their trip back to the surface. Their long stay at more than 400 feet meant four days of decompression, but it had been worth it. The two men had completed the longest dive to such a depth ever made up to that time.

## AN UNDERWATER VEHICLE TAKES SHAPE

With the invention and successful testing of these devices, the dream of living and working underwater had become a reality. During the time that Link was experimenting with SPID (and later IGLOO, another deep-sea "house"), other individuals were also making progress in this area of research. Jacques

Cousteau, for example, attracted worldwide attention with his series of Continental Shelf Stations, or Conshelfs. Conshelf I, located off the coast of France in 33 feet (10 meters) of water, housed two men for seven days. Conshelf II sat on the floor of the Red Sea in 36 feet (11 meters) of water. It was home to five men for an entire month. Link's and Cousteau's pioneering work in this area paved the way for the development of other underwater dwellings, such as the U.S. Navy's *Sealab*, in the years that followed.

Being able to live and work underwater for long periods added a new dimension to ocean exploration. But there were also drawbacks to this new technology. Remaining submerged for extended periods of time presented special risks for divers. Their bodies became accustomed to the pressures of deep water. If their life support systems failed, or if they were injured, they could not simply swim to the surface for help. Without proper decompression, such a rapid return to the surface would be fatal. The undersea habitats were also expensive to build and operate. More importantly, at least to explorers such as Link, divers

**This painting shows *Sealab III*, one of the underwater dwellings developed by the U. S. Navy in the 1960s.**

who ventured out of their undersea dwellings into the surrounding water were still quite limited in the distance they could travel through the marine world.

Something else was needed—a device that would allow people much more freedom to explore the undersea environment. Faced with this new challenge, Link and other researchers went to work designing underwater vehicles that could be piloted from place to place beneath the sea.

This was the beginning of the era of manned research submersibles. A submersible is similar to a submarine in that it can travel freely underwater. Submersibles are usually quite a bit smaller and lighter than submarines, however, and most are built to withstand the enormous pressures of deep dives. Submersibles are highly maneuverable vessels designed primarily for scientific research. Another important difference between submersibles and submarines is that submersibles are usually transported by some kind of mother ship to a particular area, launched, and then later recovered.

During the late 1950s and early 1960s, research submersibles of various types were being built all around the world. Cousteau launched one of the first successful submersibles in 1959. His *Diving Saucer* was a lightweight, saucer-shaped vessel that could carry two people down to 1,000 feet (305 meters). Perhaps the best-known American submersible, the chubby yellow *Alvin*, was built in 1964 by the Woods Hole Oceanographic Institution in Massachusetts. *Alvin*'s hull is made of titanium, and it carries a three-person research team. In 1966, *Alvin* and another U.S. submersible called *Aluminaut* (a cylindrical sub made of aluminum) worked together to locate an atomic bomb that had been lost in 2,500 feet

(762 meters) of water off the coast of Spain. During these years, a number of other countries, including Great Britain, Japan, West Germany, and the Soviet Union, each developed submersibles to carry researchers to unexplored areas of the sea.

In almost all of the submersibles being built at this time, the compartment in which scientists and crew members rode remained at surface pressure while the sub was underwater. This meant that no one could leave the vessel during a dive. The research submersible that Ed Link planned, however, had a special feature that set it apart from most other subs. Link's underwater craft was to have two separate, and very different, compartments.

The forward compartment would house the pilot and one observer. A hatch in the rear of this compartment would open into a second compartment that would also accommodate two people. Like the submersible decompression chamber, this rear compartment could be pressurized so that its interior pressure was the same as the pressure outside at any given depth. From this **"lock-out" compartment**, divers could emerge to

**Ed Link *(right)* supervises the launching of his submersible *Deep Diver* in 1967.**

explore the sea. After returning to the compartment, they would seal themselves in and begin decompression. Eventually the sub would return to its mother ship, where the divers would complete the decompression process.

In 1967, Link's plans for an underwater vehicle materialized in the form of *Deep Diver*, a compact, bright yellow submersible. *Deep Diver* was able to descend to a maximum depth of 1,250 feet (381 meters). Over the next two years, this small sub transported oceanographers, photographers, and marine researchers of all kinds to many areas of the ocean off the east coast of Florida and around the Bahama Islands.

*Deep Diver* proved to be a very valuable tool for marine science and ocean exploration. With it, scientists were able to travel to never-before-explored waters. From the lock-out compartment, they could actually leave the sub to swim among the sea's inhabitants, observe them in their natural environment, and conduct all sorts of underwater experiments. Marine plants and animals also could be collected and brought to the surface in near-perfect condition for further study.

## JOHNSON-SEA-LINK: A BUBBLE IN THE SEA

To many researchers, *Deep Diver* seemed an ideal vessel for certain kinds of underwater studies. Yet Ed Link was to return to the drawing board one more time. The result of his creative efforts was a new submersible that had many of the same features as *Deep Diver* but looked startlingly different. The smooth yellow hull had been replaced by an angular aluminum framework to which batteries, ballast tanks, propellers, gas storage tanks, underwater lights, and other types of equipment were attached. But the most interesting feature was in front—the forward compartment was a large, transparent bubble, or **sphere**. The two people riding in this unique compartment would be able to see in almost every direction! At that time, no other research submersible offered scientists and explorers such a panoramic view of the marine environment.

The gleaming aluminum submersible was launched on January 29, 1971. Christened the *Johnson-Sea-Link* in recognition of Link's longtime friend and financial partner, J. Seward Johnson, Sr., the new sub was ready to begin its career in the scientific exploration of the oceans.

Over the next few years, the *Johnson-Sea-Link* proved to be an enormous success as an underwater research tool. Although small changes were made in the sub's design and additional equipment was regularly added to its aluminum framework, Ed Link was pleased with

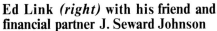
**Ed Link** *(right)* **with his friend and financial partner J. Seward Johnson**

the unique and very useful vessel he had made. So much so, in fact, that he had a second one built, and in 1975 *Johnson-Sea-Link II* took its place alongside the original submersible.

Ed Link never lost his interest in the deep sea or in creating the means to explore it. When he died in 1981, he left behind a remarkable list of accomplishments. Were Link alive today, he would undoubtedly be proud of the fact that nearly two decades after the launching of the first "bubble sub," the *J-S-L* submersibles are still hard at work helping us learn more about the ocean, the last great frontier on earth.

The *Johnson-Sea-Link II* submersible

# 3

# *EXPLORING A NEW WORLD*

With the development of submersibles and other underwater devices in the 1960s and 1970s, the ocean had suddenly become much more accessible. Ed Link and Seward Johnson were concerned about the way in which future exploration of the ocean would be conducted. They realized that the time was right and the technology available to begin tapping the ocean's enormous resources. The two men wanted to be sure, however, that the fragile marine environment would be protected and preserved. To help achieve this goal, they founded what is now Harbor Branch Oceanographic Institution (HBOI) in 1971, the same year that *Johnson-Sea-Link I* was launched.

## HARBOR BRANCH OCEANOGRAPHIC INSTITUTION

Harbor Branch Oceanographic Institution is a private, non-profit center for research in marine biology and ocean engineering located just north of Fort Pierce, Florida. It is built on 460 acres (186.2 hectares) of land surrounding Link Port, a deep channel that opens onto eastern Florida's coastal waterway. Harbor Branch is the "home base" for the two *Johnson-Sea-Link* submersibles. At the head of the channel stands a huge engineering building that houses the subs when they are not being used at sea. This

**A view of Harbor Branch Oceanographic Institution from the air. In the foreground is the engineering building, which stands at the head of a deep channel opening onto Florida's coastal waterway. Beyond the islands at the top of the photograph is the Atlantic Ocean.**

is also where engineers, draftsmen, and technicians work year-round designing and manufacturing all sorts of specialized equipment for ocean research.

Beyond the engineering building, research laboratories line the channel. Dozens of scientists work here, investigating everything from the life histories of marine organisms to the effects of pollution in the open ocean. Visiting scientists from other institutions also come to Harbor Branch to participate in research projects going on there.

A large part of the research that is carried out at HBOI would not be possible without the *Johnson-Sea-Link* submersibles. These subs, each equipped with a remarkable set of underwater tools, are helping marine biologists, ecologists, and oceanographers to learn about the oceans. Let's take a close look at one of these research submersibles and find out why it is so valuable to those who study and explore the ocean frontier.

# THE *JOHNSON-SEA-LINK* SUBMERSIBLE

The *Johnson-Sea-Link* submersible is a rather strange-looking vessel. From the outside, it resembles a cross between a helicopter and a small space ship. But its design makes it ideally suited for underwater exploration.

## THE TWO COMPARTMENTS

The first thing that catches your eye when you see the sub is the glistening Plexiglas sphere at the front. The sphere is 6 feet (1.8 meters) in diameter, and its transparent walls are 4 inches (10.2 centimeters) thick. It is not made out of a single piece of Plexiglas, but of 12 five-sided pieces that are fused together. The

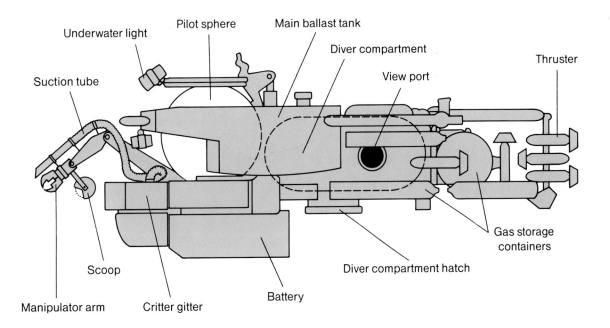

**A simplified drawing of a *J-S-L* submersible showing its main parts**

**The transparent Plexiglas sphere of the *Johnson-Sea-Link* submersible offers its occupants a panoramic view of the marine environment.**

sphere's structure makes it extremely strong. The *J-S-L* can descend to a maximum depth of 3,000 feet (914.4 meters), but the sphere alone could withstand the pressure exerted by the surrounding seawater down to 8,000 feet (2,438.4 meters).

The two people who ride in the sphere during a dive enter the transparent compartment through a circular metal hatch at its top. Once inside, they sit side by side facing forward—the pilot on the left and the scientist or observer on the right.

Directly behind the sphere is the diver, or lock-out, compartment. The diver compartment is constructed entirely of aluminum. It has only three small, circular view ports: one on each side and one in the hatch door, which is located in the floor of the compartment. The diver compartment is approximately 8 feet (2.4 meters) long, but the interior diameter is less than 4 feet (1.2 meters), so there is little room to spare.

The diver compartment is designed to connect with a large decompression chamber that can be carried on board the sub's mother ship. Divers returning from a lock-out dive in the sub can be transferred to this chamber to complete their period of decompression in more spacious and comfortable surroundings.

In recent years, the diver compartment has come to be used primarily for scientific observation rather than lock-out diving. This is because most of the tasks that divers performed outside the submersible can now be accomplished using remotely operated tools attached to the sub. Unless a lock-out dive is planned, surface pressure is maintained in the diver compartment throughout the dive, just as it is in the sphere.

**One of the circular view ports in the diver compartment**

## MANEUVERING UNDERWATER

Out of the water, the 23-foot (7-meter) *Johnson-Sea-Link* submersible looks a bit awkward and unwieldy. With equipment projecting in all directions from its angular frame, it could hardly be described as a sleek and streamlined underwater vessel. The sub weighs over 23,000 pounds (10,432.8 kilograms), so it takes

33

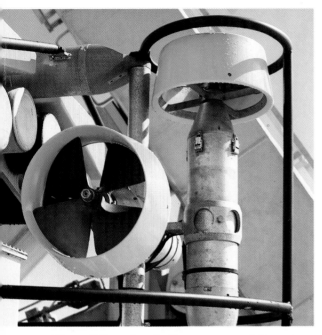

**Two of the nine battery-powered thrusters that propel the submersible through the water**

sive and easy to handle, the pilot can bring it very close to underwater objects.

By adjusting the amount of water and air in the sub's ballast tanks, the pilot controls the vessel's buoyancy in the water. Letting water into the tanks, for example, makes the sub heavier so that it descends. Forcing water out of the tanks with compressed air makes the sub lighter so that it rises. The sub can also hover virtually motionless in the water or gently come to rest on the ocean floor.

## NAVIGATION AND COMMUNICATION SYSTEMS

The *Johnson-Sea-Link*'s navigation equipment enables the pilot to determine the sub's speed, its exact position in the water, and its distance from a given point and from the bottom. An on-board sonar system produces an accurate "picture" of what lies ahead and to the sides of the sub. Electronic sensors provide information on external conditions such as the temperature of the water, how salty it is, and how much light is filtering down from above.

During a dive, the pilot communicates by radio with the sub's mother ship. Although the sphere and the diver

considerable effort and specialized equipment to lift or move it in any way.

Yet the moment the sub enters the ocean, any impression of awkwardness instantly disappears. It glides smoothly through the water, driven by nine battery-powered **thrusters** (small, motor-driven propellers) mounted at the front, sides, and rear of the vessel.

Although the *J-S-L* has a top speed underwater of less than 2 knots (2.3 miles, or 3.7 kilometers, per hour), it is very maneuverable and can travel in any direction. Because the sub is so respon-

**This underwater photo of the *J-S-L* shows the manipulator arm extended in front of the sphere. Notice the grasping jaws at the arm's tip.**

compartment are physically separate, they are linked by an intercom system. The two scientists or observers on board also wear lightweight headsets during each dive. These allow them to keep up a running conversation with each other about the things they are observing and collecting.

### RESEARCH EQUIPMENT

The *Johnson-Sea-Link I* and *II* are research submersibles. Their primary purpose is to aid scientists in learning about the oceans and, for that job, they are very well equipped.

One of the *J-S-L*'s most important pieces of equipment is a hydraulic **manipulator arm** that is mounted on the sub's front end. This giant aluminum arm extends 9 feet (2.7 meters) and is powerful enough to lift 200-pound (90.7-kilogram) objects. At the arm's free end is a pair of grasping jaws that can be rotated. These mechanical jaws are delicate enough to pick up fragile objects, yet strong enough to crush rock. In the hands of a skilled sub pilot, the mechanical arm becomes an extension of the people inside. It is used for a wide variety of underwater tasks, from breaking off pieces of coral to putting out and retrieving equipment.

In addition to the jaws, a "clam bucket" scoop and a clear plastic suction tube can be attached to the arm's end. The scoop is used to pick up all sorts of small objects, including bottom-dwelling animals such as sea stars and sea urchins. The suction tube is designed for collecting relatively delicate marine plants and animals. It is used as a sort of underwater vacuum cleaner to suck up organisms suspended in the water, clinging to the sides of rock, or crawling along the sea floor.

The suction tube is connected to a flexible plastic hose that leads to what is fondly known as the "**critter gitter**." This is a set of 12 clear Plexiglas containers fitted into a rotating platform that is mounted to the front of the sub. As each container is filled, an empty one is moved

**A close-up view of the manipulator arm's jaws** *(left)* **and the suction tube** *(right)*

**These plastic containers are part of the submersible's critter gitter. Here the "critters" are sea urchins collected during a dive to the ocean bottom.**

into place under the end of the hose. Organisms picked up with the scoop can also be deposited into containers on the critter gitter.

As researchers explore the ocean depths, they are careful to record everything they see and do. Several kinds of still cameras mounted outside the submersible are equipped to take hundreds of pictures during each dive. One of these camera systems is used primarily for photographing objects on the ocean bottom. This camera, located on the front of the sub, is focused using laser beams. Two lasers (mounted on either side of the sphere) each project a beam of colored light onto the surface of the object being photographed. By adjusting the position of the sub in the water, the pilot works to get the two beams of light to merge into a single brilliant dot. When this happens, the object is in focus for the camera, and one of the sphere's occupants quickly snaps the picture.

Color videos of the underwater world are also made using a video camera mounted outside the sub. Inside the sphere, the occupants can see exactly what the camera is recording on a color monitor located between the pilot and observer. There is also a tiny video monitor in the diver compartment so that the two people riding in the rear can see what is happening directly in front of the submersible.

## SAFETY FEATURES AND LIFE SUPPORT SYSTEMS

Pilots, divers, and scientists are well aware that deep-sea research is a risky business. Yet in all the years that the *Johnson-Sea-Link* submersibles have

been in operation, there has been only one accident that ended in tragedy. This occurred early in the original *J-S-L*'s underwater career.

On June 17, 1973, *Johnson-Sea-Link I* was diving in several hundred feet of water off the coast of Key West, Florida. Those aboard were studying fish populations near the wreck of a sunken navy destroyer. As the sub maneuvered around the wreck, it was caught in a very strong current. Before the pilot could respond, the sub was swept into a mass of tangled underwater cables from the sunken ship.

The submersible was trapped. The pilot tried repeatedly, but could not free it from the cables. At the surface, those aboard *Sea Diver* radioed for help from the Coast Guard and the Navy. Several rescue attempts were made, some by helmeted divers, others by small submersibles with mechanical arms. One attempt after another failed. Meanwhile, the submersible's occupants were running out of air.

Finally, nearly 30 hours after the sub had become entangled, a grappling hook rigged together with a closed-circuit television was lowered from a ship at the surface. Carefully the hook was brought closer and closer to the sub, until suddenly it snagged the helpless vessel. Within minutes, *Johnson-Sea-Link I* was hauled to the surface. The two men up front in the sphere were weak, but still alive. However, the two in the rear compartment—Albert Stover and E. Clayton Link (Ed Link's 31-year-old son)—had not been so fortunate. The build-up of carbon dioxide in the compartment, combined with the body-numbing cold, had cost the men their lives.

As a result of this terrible tragedy, Ed Link drew up plans for a new device that could be used to rescue disabled submersibles. It was a remotely controlled, unmanned vehicle called **CORD**, short for **Cabled Observation and Rescue Device**. CORD was equipped with TV cameras, sonar, cable cutters, and other equipment.

In recent years, other remotely operated vehicles, or **ROVs**, have replaced CORD at Harbor Branch. HYSUB 40, for example, is a ROV that can be used to a depth of 3,280 feet (1,000 meters) and can remain underwater almost indefinitely. It is equipped with two small manipulator arms, video cameras, and

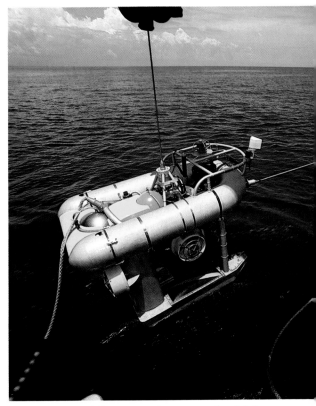

**CORD, one of the remotely operated vehicles developed for use in rescuing disabled submersibles**

39

powerful underwater lights. ROVs such as HYSUB 40 are operated from on board a surface support vessel. At Harbor Branch, Ed Link's *Sea Diver* has been fitted out for this purpose.

Although there will always be risks involved in diving thousands of feet below the surface, Harbor Branch engineers work to make sure that each dive will be as safe, if not safer, than the one before. Dozens of features are built into the sub to reduce potential risks to the vessel and its passengers. For example, all the viewing ports, hatches, and places where lines and wires enter the compartments are designed to seal tighter and tighter as the submersible descends deeper and deeper. Any new equipment designed for use with the sub underwater is rigorously tested before actually being used on a dive. And there are backup systems for everything!

Furthermore, the sub's "open" framework makes it easier to spot small problems that might eventually develop into something more serious. If a piece of the sub's equipment is found to be faulty before or after a dive, it can usually be serviced without taking apart the entire vessel to get at the problem. A damaged

thruster, for example, can be removed and replaced with a new one in minutes. Skilled sub technicians make the job look almost as easy as changing a car's flat tire.

Once underwater, however, the little submersible is really on its own. But in the event that something totally unexpected should happen to the *Johnson-Sea-Link* during a dive, there are a number of ways that the sub can surface, even if all power is lost. First the pilot can release compressed air into the water-filled variable ballast tanks. As the water is driven out of the tanks by the air, the sub will become lighter and more buoyant. Within seconds, it should begin to rise.

If this doesn't work, the pilot would next release compressed air into the main ballast tanks. If the sub still won't rise, even after the main ballast tanks are completely emptied of water ("blown dry"), then another step can be taken. The large and heavy battery pod located underneath the sphere can be released. At this point, unless the submersible is caught on something, it should literally bob to the surface.

Provisions for a long underwater stay

***Johnson-Sea-Link II*** **in the shop at Harbor Branch. The two submersibles are serviced and inspected regularly to make sure that they are in perfect working condition.**

have also been greatly improved since the 1973 tragedy. If a *J-S-L* submersible should ever become trapped, it carries enough oxygen, water, and supplies so that the four people inside could survive for five days, if necessary, while help was on the way.

The two *Johnson-Sea-Link* submersibles are serviced regularly to make sure

that they stay in good shape. Sometimes the subs are taken completely apart, piece by piece, in the workshops at Harbor Branch. Each part is inspected, refurbished, or replaced. When the vessels are reassembled, they are rigorously tested. They must be in perfect working condition to carry out the demanding task of underwater exploration.

# THE MOTHER SHIPS

The *Johnson-Sea-Link* submersibles can do many remarkable things, but without their mother ships, or **surface support vessels**, they could not get very far. Harbor Branch has two such ships that transport the submersibles from place to place. They also launch the subs and, after each dive, retrieve them from the water.

The largest ship in Harbor Branch's fleet is the 176-foot (53.6-meter) R/V *Seward Johnson*. (The letters R/V stand for "research vessel.") This ship was designed and built solely for marine research using the *Johnson-Sea-Link* submersibles. In addition to ship and sub crews, the *Seward Johnson* can accommodate nearly a dozen scientists. This ship is a floating laboratory where researchers can study, photograph, and preserve the collection of specimens that the sub brings up from each underwater voyage.

Sitting atop the *Seward Johnson*'s rear deck is a massive cranelike device shaped somewhat like a giant capital **A**. Descending from the top cross beam of the **A** is a telescoping structure that locks on to the top of the submersible. This A-frame **submersible handling system** is used to launch and recover the sub.

To launch the submersible, the telescoping part of the A-frame first lifts the sub up out of a padded metal cradle that is mounted on the ship's rear deck. Then the entire A-frame slowly tilts out over the back of the ship. Once over the water, the sub is gently lowered down into the sea on a thick cable. When the cable is released, the sub is free in the water.

To retrieve the sub from the water, the process is reversed. The sub is hooked back up to the thick cable and lifted straight up out of the water. As the A-frame returns to an upright position, the sub is brought back on deck once again and replaced in its cradle.

During the time that the sub is suspended in the air, it is held securely by the handling system so that it cannot swing back and forth. This is critically important because a swinging sub might crash into part of the ship as it was being moved. The A-frame handling system is so efficient that a *Johnson-Sea-Link* can be launched or recovered in a matter of minutes.

On the rear deck of its mother ship, *Johnson-Sea-Link II* sits beneath the A-frame submersible handling system.

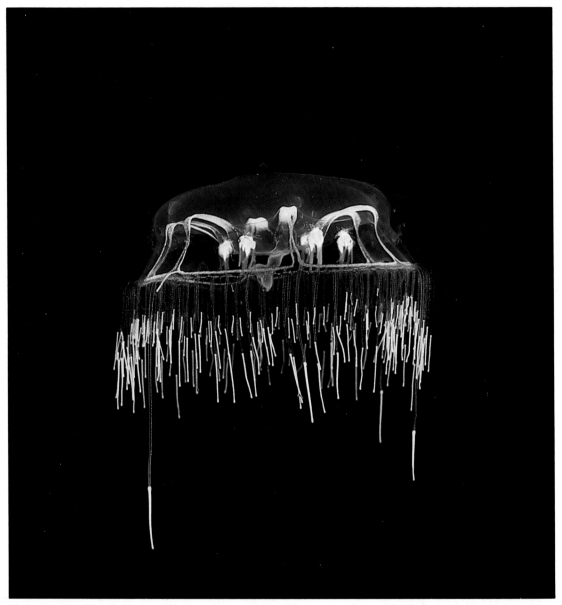

Scientists at Harbor Branch are using submersibles to study zooplankton, small animals that float and drift through the ocean. This photograph shows *Halicreas minimum*, a jellyfishlike member of the plankton. *Halicreas* is a predator that captures and kills prey with its tentacles.

# 4

# *A JOURNEY TO THE OCEAN FLOOR*

Every year, scientists from Harbor Branch, along with visiting researchers from other places, use the *Johnson-Sea-Link* submersibles in marine research. The year's work is divided into a series of cruises. Each cruise involves transporting one of the submersibles to a particular destination where scientists wish to explore the ocean, collect marine organisms, and conduct experiments. A scientific cruise may last only a few days or as long as several months. It all depends on what the scientists are working on and where they need to go to carry out their investigations.

In the past few years, for instance, some of the marine biologists from Harbor Branch have used the *Johnson-Sea-Link* to observe and collect **zooplankton** along the eastern coast of the United States. Zooplankton is a general name for the millions of small animals that drift through the seas. Some kinds of zooplankton live near the surface; others inhabit the ocean depths. Many of these animals are essentially transparent and thus almost impossible to see in the water. Their bodies tend to be very fragile and are easily damaged when collected using standard methods such as nets made of fine mesh. As a result, few species of zooplankton have been studied in detail, and relatively little is known about the great majority.

But with the aid of a submersible, researchers can come face to face with planktonic animals. An elaborate camera system, complete with four strobe lights, can be mounted to the front of the *Johnson-Sea-Link*s and used to capture these spectacular and often bizarre creatures on film. The submersibles can also be equipped with special canisters in which zooplankton are collected unharmed and carried up to the surface in perfect condition for study.

Other biologists at Harbor Branch are trying to piece together the complex life cycles of some deep-water animals that live on the sea bottom. Certain species of **echinoderms** (a group of marine animals that includes sea urchins, brittle stars, sea stars, sea cucumbers, and sea lilies) are found only at great depths, in several thousand feet of water. It is nearly impossible to collect and study such animals from the surface. In the *J-S-L*, however, scientists have made many trips to the ocean floor in places such as the Gulf of Mexico and the Caribbean to collect these bottom-dwellers and bring them to the surface for detailed study.

Potentially useful chemical substances can be extracted from many marine

HBOI researchers are also studying deep-sea echinoderms like this sea urchin *(Tretocidaris bartletti)*.

organisms. Another area of challenging research being carried out at HBOI and elsewhere involves experimenting with such "extracts" in the hope of developing new drugs. In 1986, *Johnson-Sea-Link II* was transported to the Galápagos Islands off the western coast of South America. For several months, scientists used the submersible to collect sponges, marine plants, and certain kinds of marine animals to use in these experiments.

# PREPARING FOR A SCIENTIFIC CRUISE

Scientific cruises involving the *Johnson-Sea-Link* submersibles are planned many months in advance. One of the first things researchers must do is obtain funding for their marine expeditions and for the experimental work they want to do during and after each cruise. Submersibles are very expensive to operate. It costs approximately $13,000 per day to operate one of the *J-S-L*s, complete with its mother ship and crews. Harbor Branch pays for part of the cost of its scientists' marine research. Additional funding also comes from outside sources, such as the National Science Foundation and the National Oceanic and Atmospheric Administration.

Researchers preparing for a cruise often need special pieces of equipment to conduct underwater experiments. With the help of engineers and craftsmen at Harbor Branch, the specific tools required for their research will be designed, manufactured, and, if necessary, attached to the submersible.

During this time, the sub's mother ship is also being made ready for the voyage. Because it may have to spend weeks at sea, far from any port, the vessel must be as self-sufficient as possible. This means that everything from food and medical supplies to spare parts for the submersible must be carried on board.

When all the preparations are completed, the sub is loaded aboard the mother ship. With fuel tanks full and supplies securely stowed, the R/V *Seward Johnson* is ready to leave on her first scientific cruise of the season. Why don't we climb aboard?

# DESTINATION: THE BAHAMAS

Our cruise begins as the *Seward Johnson*, with *Johnson-Sea-Link II* secured to her rear deck, steams out of Link Port at high tide. Turning south, the big, white ship travels slowly down the coastal waterway for several miles and then heads out into the open ocean.

Once at sea, she cruises along at a speed of 12 knots (about 13.8 miles, or 22.2 kilometers, per hour). As the bow of the ship slices through the clear blue water, flying fish soar out of the waves, skimming over the surface as they "fly."

**The *Seward Johnson*, with *Johnson-Sea-Link II* secured to her rear deck, cruises through the blue waters of the Atlantic Ocean.**

Occasionally a sea turtle surfaces, raising its head to gaze briefly at the passing ship before slipping back into the sea.

After a 16-hour trip that carries the ship southeast across the Gulf Stream, the *Seward Johnson* arrives at its destination: Nassau, a colorful port city on New Providence Island in the Bahamas. After clearing customs and obtaining permission from the Bahamian govern-

ment to dive in its waters, the *Seward Johnson* anchors in Nassau's busy harbor. Here the ship and her crew wait to be joined by the group of scientists who will be using the *Johnson-Sea-Link* to conduct a variety of deep-sea investigations on this trip.

The members of the science party arrive the following morning by plane. Two are marine biologists from HBOI;

they will be in charge of the research that takes place during this cruise. They are interested in finding out as much as possible about the breeding habits and early development of several kinds of deep-sea echinoderms, particularly sea urchins. Two graduate students, one from Harbor Branch and one from Florida State University, will assist them in their studies. Also from HBOI is an **electron microscopy** specialist. She will be responsible for preserving parts of specimens so that they can be studied in minute detail using electron microscopes back at Harbor Branch.

Two visiting scientists from the Smithsonian Institution are along to study zooplankton and marine worms. There is also a marine biologist from Seattle, Washington, who hopes to collect and photograph several kinds of rare marine snails. Another visiting biologist, an expert on deep-sea echinoderms, has come all the way from Great Britain to be a part of this cruise. Finally, a visiting observer (the author) completes the group. She will be recording in detail the experiences and discoveries that occur during the next two weeks.

Once on board, the researchers quickly settle in. This ship will be their home and workplace for the next two weeks. Each person is first assigned to a cabin located on either the ship's lower or middle decks. There is not much space available so two or three people must share each small room. The cabins are equipped with bunks, a sink, a small bathroom with toilet and shower, and storage cabinets for stowing clothing and personal items.

Next, the members of the group carefully unpack microscopes and other pieces of delicate equipment they have brought and fasten them to workbenches in the ship's laboratories. (Everything on board must be secured well enough to survive a rough ride in case of bad weather.) There are two labs on board. The "wet" lab has sinks and workbenches along its walls. In the center of the room, there is also a large table with a drain at one end. Most of the messy work of sorting, washing, and dissecting organisms brought up by the submersible will take place in the wet lab.

The "dry" lab will be used for microscope work and for conducting experiments. This lab also contains sinks and workbenches, as well as incubators, refrigerators, and freezers for keeping animals and experiments at specific temperatures. One corner of the dry lab is devoted to video equipment, while across the room is an area used for developing film from the sub's cameras.

As the scientists tend to their various tasks, the ship leaves the security of Nassau's harbor and heads out to sea. Her course is set for a point just off the southwestern tip of New Providence Island. This is the site where the first dive of the cruise will take place.

It will take four or five hours for the *Seward Johnson* to reach the dive site. During the trip, the scientists finish their unpacking and then join some of the crew members for lunch in the ship's galley. There are places for 14 people to sit and eat in the galley, but since there are 25 people on board (8 ship crew, 7 sub crew, and 10 scientists), not everyone is able to eat at the same time. The ship's cook serves three hearty meals each day. The evening's menu often includes fresh seafood; shrimp, scallops, and deep-fried squid are especially popular. Those on board can also help themselves to fruit, ice cream, hot and cold drinks, and other snacks at any time, day or night.

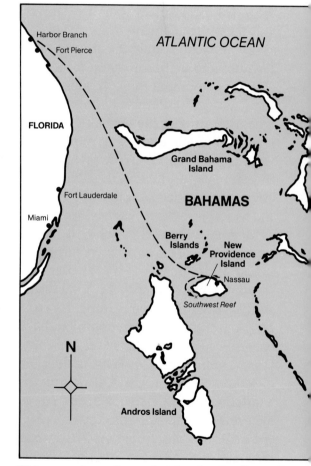

This map shows the route taken by the *Seward Johnson* to the port of Nassau and to the site of the first dive.

# A DIVE TO 2,400 FEET

Shortly after lunch, the ship arrives at its destination. The weather is calm and clear—perfect for launching the submersible. The captain of the *Seward Johnson* takes a reading of the water's depth. The instruments tell him that directly beneath the ship the sea floor is 2,400 feet (731.5 meters) down. The captain slows the engines, and the ship comes to a stop in the water.

Moments later, a loud bell sounds three times. This is a signal to all on board that the ship has arrived "on station" and that the sub is ready to dive. At the sound of the bell, everyone hurries to the ship's briefing room. Gathered around one of the submersible's pilots, they listen carefully as he reviews the dive plan. One of the scientists from the Smithsonian will sit up front in the sphere with the pilot. Behind them, in the diver compartment, will be a marine biologist from Harbor Branch, accompanied by an experienced member of the sub crew. The goal of this dive is to collect several species of sea urchins and to record the sea floor and its inhabitants on film.

When the meeting is over, the members of the expedition make their way to the rear deck. The two people who will be riding in the diver compartment scramble up through the hatch at its bottom. It is rather cramped inside; switches, gauges, and life support devices line the curved walls of the tubular chamber. The scientist sits in the rear end of the compartment, nearest the two small view ports that look out on either side of the sub.

Meanwhile, at the sub's front end, the pilot has climbed a ladder to the hatch at the top of the sphere and eased himself down into his seat inside. He now begins a final check of the sub's equipment. First, he determines that the communication, navigation, and life-support systems are all operating normally. Next, he tests the mechanical arm, the critter gitter, the cameras, and the underwater lights. When he is satisfied, he signals to a crew member poised at the top of the ladder, who in turn tells the scientist waiting below that it is all right for her to climb up and board the sub.

Wriggling down through the small hatch opening, the scientist settles into her seat next to the pilot. The sphere

**A marine biologist enters the submersible's sphere through the hatch at the top.**

more instruments, dials, and gauges, as well as the sub-to-ship radio.

The crew member standing atop the sub closes the sphere's metal hatch with a gentle thud. Directly beneath the closed hatch, the pilot attaches a large metal cylinder called a **scrubber**. During the dive, air inside the sphere is constantly circulated through this device. Chemicals contained inside the scrubber remove carbon dioxide exhaled by the sphere's occupants and, in so doing, keep the air pure and fresh. The diver compartment is equipped with a similar scrubber.

As the scientist dons her headset, the pilot speaks briefly over his radio to the captain of the *Seward Johnson*. Everything is ready to go. On deck, an engineer steps up to the controls that operate the A-frame handling system. Crew members release the heavy metal fasteners that hold the sub firmly in its cradle on the ship's deck. Now those on deck back away from the sub as the signal is given— "Prepare to dive!"

Overhead, the giant A-frame comes to life. With the *Johnson-Sea-Link* held tightly in its mechanical grip, the telescoping portion of the A-frame lifts the sub several feet straight up off the deck.

seems quite roomy even though it is packed with equipment of all kinds. Just overhead, an instrument panel curves around the front of the sphere's transparent wall. Between the two seats are other devices, including a sonar screen, the video monitor, and the controls for the video camera. On the pilot's left are

As the entire structure tilts slowly backward, the submersible is moved out over the water. For a few seconds, the sub hangs suspended over the rolling sea. Then it is lowered down into the waves with a gentle splash. Finally, the thick cable locked into the top of the submersible is released, along with a towline attached to the sub's front end. The *Johnson-Sea-Link* is on its own.

Very slowly, the sub backs away from the stern of its mother ship. For a few moments, it bobs on the surface. Then, as the pilot allows seawater to flow into the ballast tanks, the submersible becomes heavier and begins to sink. With a flurry of bubbles, it disappears beneath the waves.

For those inside the submersible, the world outside is momentarily hidden by the mass of swirling bubbles. But as the sub continues to descend, the view abruptly clears to reveal bright blue-green water stretching out in all directions. Underwater, the transparent walls of the sphere seem to disappear and the riders up front have the feeling that they are part of the ocean around them.

The sub steadily descends, dropping at a rate of 100 feet (30.5 meters) per

**Crew members watch while the *J-S-L* is launched from the rear deck of the mother ship.**

minute. At 65 feet, a big jellyfish pulses past the sub, its long tentacles trailing below its umbrellalike body. At 150 feet, the pilot spots a ghostly shadow lurking in the distance—a curious shark, perhaps? The minutes tick by. As the light from above gradually fades, the color of the water changes from blue-green to bright blue, and then to a deep royal blue. At this point, the pilot turns on the underwater lights, transforming the sub into a tiny beacon of light in a deep blue sea.

750, 1,000, 1,250 feet... the submersible continues its journey downward. At 1,500 feet (457.2 meters), there is nothing but inky blackness on all sides. This is a forbidding world of eternal night. Yet, every so often, brief flashes of colored light—some blue, some green—appear here and there in the dark water around the sub.

This eerie, twinkling light display is the work of animals that live at these great depths. A great variety of deep-sea

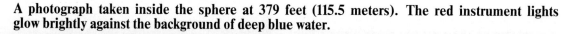

**A photograph taken inside the sphere at 379 feet (115.5 meters). The red instrument lights glow brightly against the background of deep blue water.**

creatures are **bioluminescent**; that is, they can produce their own light. Up to 80 percent of the animals in the ocean, including jellyfish, squid, shrimp, and many kinds of fish, are bioluminescent.

Different deep-sea animals have different ways of producing light. Some have light-producing bacteria in their body tissues. Others discharge clouds of a luminous secretion into the water. Many have tiny light organs called **photophores** located in various places on their bodies.

The gentle passage of the submersible disturbs the light-producing inhabitants of the deep, and they respond with flashes of light. Sometimes the pilot will use the sub's underwater lights to "talk" to these animals. A series of brilliant flashes from the sub will often bring answering flashes from the remarkable assortment of organisms swimming and drifting past.

At 2,200 feet (670.6 meters) down, the pilot announces that the sea floor is just minutes away. He slows the sub's descent by adjusting the amount of air in the ballast tanks. A hush falls over the occupants of the sub as they strain to catch the first glimpse of their destination. Then, illuminated by the *Johnson-*

Like many bioluminescent animals, this deep-sea squid *(Chiroteuthis veranyi)* produces light by means of tiny organs called photophores.

The submersible's underwater lights illuminate two inhabitants of the ocean floor. These strange animals are sea lilies, echinoderms whose feathery arms are supported by jointed stalks.

*Sea-Link*'s underwater lights, an expanse of pale sand suddenly comes into view. The instruments indicate a depth of 2,397 feet (730.6 meters).

The pilot gradually brings the sub to within a few feet of the bottom. Then he engages the rear thrusters, which hum softly as they propel the submersible forward. Cruising slowly over the eerie landscape, the sub's occupants begin their journey through an alien world whose inhabitants dwell in continual darkness.

Small crabs dot the sea floor, their pincers outstretched in front of them. Red-and-white-striped shrimp lie on the cream-colored sediments with their long, jointed legs spread out around them like fans. Slender-armed brittle stars creep along the bottom, leaving a faint trail in the soft sand. A bright orange fish swims slowly into the pool of light created by the sub, its enormous eyes fixed on this strange vessel that has invaded its territory.

After traveling for several minutes, the sub's powerful lights reveal a rocky outcropping up ahead. Perched atop a huge silt-covered boulder are several sea urchins; their dome-shaped bodies are covered with very short (½-inch or 1.3 cm) spines. The scientist in the sphere trains the sub's video camera on the boulder so her companion can see the urchins on the small video monitor in the diver compartment. After a moment's discussion over their headsets, the scientists decide to collect several of these urchins. They belong to one of the species the researchers on this cruise wish to study.

First, however, photographs are taken of the urchins in their natural surroundings. The twin beams of light from the laser-focused camera sparkle as they flicker across the body of one of the urchins. The pilot carefully positions the sub so that the two laser beams merge into a single bright spot of light. The moment they do, the scientist triggers the camera.

After several pictures have been taken, the pilot skillfully maneuvers the sub to within inches of the massive rock. As the sub hovers motionless in the water, he brings the manipulator arm into play.

These dome-shaped sea urchins *(Conolampas sigsbei)* were photographed and then collected during the dive.

Carefully he positions the scoop attached to the arm directly over one of the urchins. With a flick of the controls, the sides of the scoop close gently around the spiny animal, capturing it inside. The arm moves up and over the critter gitter, and as the scoop opens, the urchin drifts down into an empty container.

The process is repeated until half a dozen of the urchins are safely inside three of the critter-gitter containers. The scientists carefully record the numbers of the containers as well as the water temperature, depth, and any other information that may be important when studying the urchins back in the ship's laboratories.

The sub moves on, passing over the boulders and descending onto a flatter, more open area of the sea floor. Here the scientist in the diver compartment spots more urchins to the left of the sub. They are *Salenia*, small urchins with delicate, pink-and-white-striped spines about 3 inches (7.6 cm) long. These fragile creatures can be collected with the suction tube. The pilot brings the tube next to each animal and then turns on the suction. The urchins slip up the tube one after another.

As the last *Salenia* settles into the critter gitter, another sea urchin comes into view. This species is *Calocidaris micans*, a large animal with very long (8-inch or 20.3-cm) white spines that project from a pale green body. *Calocidaris* is so large that it will not fit inside the critter gitter's scoop or pass through the suction tube. To collect this species, the submersible pilot must come up with a special technique. He uses suction to hold the urchin at the end of the tube while positioning it over an empty container of the critter gitter. When he turns off the suction, *Calocidaris* gently settles inside.

Collecting the large sea urchin *Calocidaris* (above) requires a rather special technique (below).

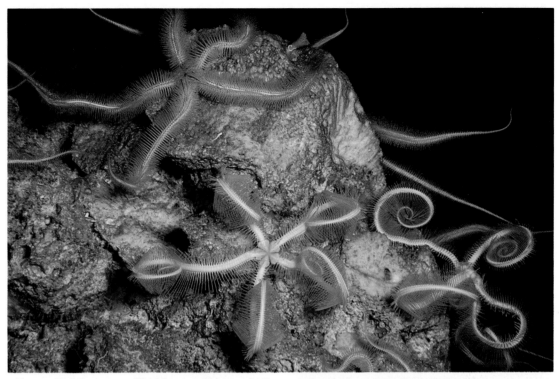

*Above:* **Brittle stars *(Ophiocamax)* perched on a rock wave their long arms as the sub passes.** *Right: Peltaster* **is a bright yellow sea star that makes its home on the ocean floor.**

For the next two and a half hours, the *Johnson-Sea-Link* carries its passengers through a world that few people have ever seen. They find several more species of sea urchins, along with clusters of brittle stars, chunky blue sea cucumbers, stalked sea anemones, and delicate sponges that look like vases made of spun glass. The beauty and variety are spellbinding.

The *J-S-L* bobs on the surface after its ascent from the sea floor.

The sphere's occupants continue to collect specimens until the containers in the critter gitter are full. They take dozens of still pictures and, with the video camera, record some of the amazing sights they witness during their stay on the bottom: rubbery sea stars that slither along the bottom at surprising speeds; a brittle star that "swims" up through the water when disturbed, waving its arms like a dancer in an underwater ballet; and a 12-foot (3.7-meter), six-gilled shark that tries to take a bite out of one of the sub's underwater lights!

## RETURNING TO THE SURFACE

The time on the bottom passes quickly. Before long the pilot announces that the sub must begin its ascent to the surface. Reluctantly, the four explorers watch as the sea floor beneath them slips back into total darkness. The sub rises steadily through the water, which gradually changes from black to dark blue to light blue and eventually to a dazzling blue-green. Finally, with a sudden burst of bubbles, the *Johnson-Sea-Link* breaks the surface and bobs up into the welcome sunshine.

During each dive, a crew member aboard the mother ship closely monitors the sub's whereabouts with sonar. For several hours, the *Seward Johnson* has maintained a position almost directly above the submersible far below. When the *J-S-L* breaks the surface, its support vessel is close by. Now the ship heads toward the waiting sub and the recovery process begins.

As the ship pulls up alongside the *Johnson-Sea-Link*, a crew member dives off the rear deck carrying the towline. With a few powerful strokes, he reaches the sub's front end and attaches the towline just above the sphere. Then he climbs up on top of the sub while the *Seward Johnson* moves slowly forward, pulling the submersible into position for recovery behind the ship.

Next the massive cable from the telescoping portion of the handling system is let down above the sub. The diver reaches up, grabs the cable, and quickly inserts its metal end into the top of the submersible. The A-frame motor hums loudly as the *Johnson-Sea-Link* is plucked from the sea and deposited safely on the deck.

**A diver attaches the handling system cable to the top of the submersible so that it can be lifted onto the mother ship.**

Scientists eagerly gather around as the critter gitter is unloaded.

When the hatches are opened, the sub's occupants emerge, enthusiastic about their voyage and anxious to examine firsthand the organisms collected on the ocean floor. As the scientists crowd around the critter gitter, the excitement of the moment shows clearly on their faces. Almost every dive brings something new to the surface.

## IN THE SHIP'S LABORATORIES

As each container filled with deep-sea organisms is unloaded from the critter gitter, it is taken directly to one of the ship's laboratories. There the specimens are washed clean of sand, identified as to species, and carefully measured. Each organism is also inspected for other creatures that may be living on or in it. Several large purple urchins that were collected on this dive, for example, have tiny shrimp hiding among their spines. The shrimp match the color of their hosts almost exactly!

Following this initial examination, the scientists use some of the organisms in a series of experiments. Through their investigations, they hope to learn more about the inhabitants of the deep sea and perhaps find answers to questions such as: How long do bottom-dwelling animals live? When and how frequently do they reproduce? How do their offspring develop, and what factors in the environment affect their development? And how do these animals find other members of their own species in total darkness? Any organisms that are not used immediately in on-board experiments are carefully preserved for future studies.

By this time, the photographs taken during the dive have been developed. The researchers take a moment from their work to review the photos, as well as the videos from this dive. As the video tape rolls, everyone aboard the ship is able to relive the underwater adventure that took place only hours before.

Today's dive was very productive. It was also just the beginning of the researchers' investigations in these waters. The *Johnson-Sea-Link* is capable of making two 3½-hour dives per day. In between time, the sub's batteries are recharged, air supplies are checked and replenished, and the cameras are reloaded. Day after day during this cruise, the submersible will carry its passengers down into the darkness to explore, observe, record, and collect. By the time the *Seward Johnson* heads home to Harbor Branch, the scientists will have made more than two dozen dives in the submersible. Each dive will have provided them with one more valuable glimpse of what life is like at the bottom of the sea.

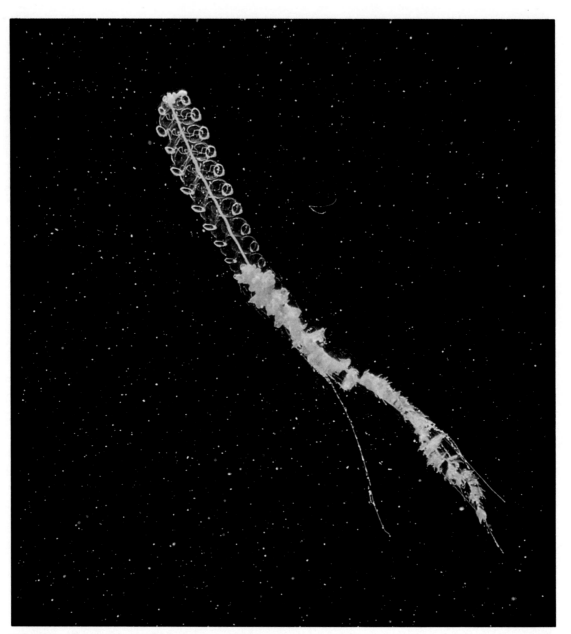

**Particles of marine snow surround a siphonophore, an animal of the zooplankton. This material is a source of nutrients for organisms living on the ocean floor.**

# 5

# *EXTENDING OUR KNOWLEDGE OF THE SEA*

The ocean still remains the least-explored part of our planet. In fact, we know less about the sea than we do about some areas of our solar system. But we are making steady progress. Each time scientists descend beneath the waves in research submersibles, they add to our knowledge of this immense body of water. And some of the things they are discovering are changing the way people think about the marine environment and the organisms that live in it.

For example, extensive studies of zooplankton in the upper 2,400 feet (731.5 meters) of the ocean have been made using the *Johnson-Sea-Link*s. By determining what zooplankton eat, where and how fast they travel, and what animals prey upon them, scientists are beginning to understand how energy, in the form of nutrients, flows through the world's oceans. Zooplankton are food for many kinds of marine predators. But interestingly, their waste products, which float down through the water as particles known as **marine snow**, are also a vital source of nutrients for organisms living far, far below on the ocean floor.

In 1983, researchers aboard *Johnson-Sea-Link I* found red algae, a kind of underwater plant, growing 877 feet (267.3 meters) below the surface. Up until that time, no one thought it was possible for plants to live at such depths where there was so little light. Since then, similar types of algae have been found growing even deeper. Algae such as these may be the deepest-growing plants on earth.

Almost every time scientists descend in the submersibles, they return having seen or collected something new. During a recent exploration of steep underwater slopes in the Bahamas, for example, scientists using the *Johnson-Sea-Link*s returned with more than 140 species of echinoderms. At least a dozen of these animals were species that had never been described before. Who knows what new and unique organisms might be illuminated by a sub's underwater lights on its next dive?

As researchers continue to explore the oceans of the world, the knowledge they gain often has important practical applications as well as being of scientific value. We now know that the waters and sea beds of oceans around the world are rich in economically important minerals such as manganese, copper, nickel, chromium, cobalt, iron, zinc, and even gold. Deep-sea photographs, for example, have revealed large deposits of manganese "nodules"—small lumps made up of manganese, copper, nickel, and cobalt—that carpet the sea floor in parts of the Pacific, Atlantic, and Indian oceans. Presently, mining for minerals in the ocean is very expensive and requires specialized underwater equipment. But as our supplies of some minerals dwindle on land, it may become necessary to mine the ocean's mineral wealth.

Oil and natural gas supplies on land are already on the decline. Today, offshore oil rigs dot the coastal waters of many countries as the search for these fuels continues in the ocean. Submersibles, both manned and unmanned, play an important role in this industry. In addition to gathering information on the environment around deep-sea oil wells, they are used to inspect miles of cables and pipelines and to make repairs on underwater equipment and structures. The quest for offshore oil has its drawbacks, such as the constant threat of oil spills at sea that contaminate water and beaches and endanger countless marine organisms. But it may have unexpected benefits as well. Researchers have found that the submerged portions of offshore oil rigs act as artificial reefs that support a diverse community of marine life.

Our expanding knowledge of undersea life also has opened up the possibility of using more of the ocean's inhabitants as sources of food. One area of current marine research is **mariculture**, the raising

**A Harbor Branch worker examines algae being grown in the large tanks behind her. HBOI researchers are studying methods of raising such "seaweeds" for use as food.**

of marine plants and animals for food. Researchers at many institutions, including Harbor Branch, are developing techniques for growing certain types of algae in the laboratory. One day these nutritious "seaweeds" may be grown in marine farms and used for both human and animal food. Other scientists are attempting to find cost-effective ways of raising large numbers of various marine animals. Fish, shrimp, clam, oyster, and lobster farms may some day help to supply food to our hungry planet.

Marine researchers hope that other "harvests" of the future will include supplies of fresh water distilled from seawater by a process known as **desalinization**. Non-polluting energy might also be produced by using the power of waves and tides.

In order to take advantage of what the ocean has to offer, however, we must make sure that it is protected from things that could alter or even destroy it. The greatest threat to the future of the ocean today is pollution. Despite some regulations, the seas are still being used as a worldwide dumping ground for garbage, sewage, pesticides, and hazardous wastes. No one really knows what the long-term effects of these pollutants will be on the marine environment. Much more research is needed to understand the problems created by this pollution and to find solutions to them before permanent damage is done to the ocean and its inhabitants.

The lure of the sea is a powerful one, just as compelling today as it was in centuries past. With the aid of versatile underwater tools such as the *Johnson-Sea-Link* submersibles, scientists will continue to explore the ocean frontier. With this continued exploration will undoubtedly come a better understanding of the marine environment. We can only hope that it will also be accompanied by a strong commitment to protect and preserve this precious natural resource for many years to come.

# GLOSSARY

**bathyscaph (BATH-ih-scaf)**—a deep-diving vessel made up of a steel sphere attached beneath a large cylindrical hull. Increasing or decreasing weight in the hull causes the vessel to move down or up. In 1960, the bathyscaph *Trieste* descended to the deepest part of the ocean, nearly 7 miles down.

**bathysphere (BATH-ih-sfeer)**—a ball-shaped steel observation chamber lowered into the ocean on a cable. In the 1930s, William Beebe and Otis Barton made several historic dives in a bathysphere that Barton had designed and built.

**bioluminescent (bi-oh-lu-mih-NES-uhnt)**—capable of producing light. Many deep-sea animals produce their own light either with special organs called photophores or by other means.

**CORD (Cabled Observation and Rescue Device)**—an unmanned, remotely controlled vehicle that can be used to rescue disabled submersibles

**"critter gitter"**—a set of Plexiglas containers fitted into a rotating platform that is mounted on the front of a *Johnson-Sea-Link* submersible and used to store organisms collected during a dive

**decompression**—the decrease in pressure experienced by a diver when returning to the surface. Too rapid an ascent can cause a dangerous condition known as the bends.

**desalinization (dee-sal-ih-nih-ZAY-shun)**—the removal of salt from seawater to produce fresh water

**echinoderms (ih-KI-nih-derms)**—sea stars, sea urchins, and other spiny-skinned marine animals that scientists include in the phylum Echinodermata

**electron microscopy (my-KROS-kih-pee)**—the study of objects using an electron microscope

**lock-out compartment**—the rear, or diver, compartment of a *J-S-L* submersible. This compartment was originally designed for use by divers who would leave and return to the sub while it was underwater. Today it is used primarily for observation.

**manipulator arm**—a mechanical arm mounted on the front end of a *J-S-L* submersible. Operated by the sub pilot, the arm is used to perform a variety of underwater tasks.

**mariculture (MAR-ih-kul-chur)**—the raising of marine plants and animals for food

**marine snow**—a collection of particles made up of the waste products of zooplankton. This material provides food for organisms living on the ocean floor.

**photophores (FOT-uh-fores)**—small light-producing organs that are found in a variety of deep-sea animals

**ROVs (Remotely Operated Vehicles)**—unmanned, remotely controlled underwater devices operated from a ship at the surface and equipped to perform a variety of underwater tasks

**scrubber**—a device containing chemicals that purify the air inside a submersible

**sphere**—the front compartment of a *J-S-L* submersible, a transparent bubble made out of Plexiglas

**SPID (Submersible Portable Inflatable Dwelling)**—an underwater "house" invented by Edwin Link in the 1960s. SPID consisted of an iron frame attached to a large balloon that provided living space for divers exploring the ocean floor.

**submersible**—a small, maneuverable diving vessel designed for scientific research in the deep ocean. Submersibles are usually transported, launched, and recovered by a mother ship.

**submersible decompression chamber (SDC)**—a cylindrical chamber designed by Edwin Link to carry divers to the ocean floor and back. The pressure inside the chamber could be adjusted to match the surrounding pressure. This allowed divers to leave the SDC while underwater and to undergo decompression inside the chamber when the dive was over.

**submersible handling system**—a large crane-like device on the deck of a mother ship used to launch a submersible and to recover it after a dive

**surface support vessels**—the large mother ships that transport, launch, and recover submersibles. Surface support vessels also contain living quarters for scientists and crew members and laboratories where scientific work is done.

**thrusters**—small, motor-driven propellers that move the *J-S-L* subs through the water

**zooplankton (zo-oh-PLANK-tuhn)**—the mass of tiny animals that drift through the ocean. Zooplankton includes some types of one-celled organisms, an enormous collection of small animals, and the eggs and larvae of larger creatures such as echinoderms, crabs, and fish. It is an important source of food for many kinds of marine life.

# INDEX

**ACKNOWLEDGMENTS** The photographs in this book are reproduced through the courtesy of: pp. 6, 18, Marty Snyderman; p. 8, Carter M. Ayers; pp. 9, 27, 30, 35, 39, 41, 48, 67, 68, Harbor Branch Oceanographic Institution, Inc.; p. 10, Metropolitan Museum of Art, Gift of Alexander Smith Cockran, 1913; p. 12, Library of Congress; p. 14, UPI/Bettmann Newsphotos; p. 16, Luis Marden, National Geographic Society © 1960; p. 21, Bates Littlehales, National Geographic Society © 1968; p. 23, United States Navy; p. 25, Bates Littlehales, National Geographic Society © 1963; pp. 28, 32, 33, 34, 36, 37, 43, 53, 54, 58 (bottom), 60, 61, 62, Rebecca L. Johnson; pp. 44, 52, 55, Pamela Blades-Eckelbarger, Harbor Branch Oceanographic Institution, Inc.; pp. 46, 57, 58 (top), 59, Department of Larval Ecology, Harbor Branch Oceanographic Institution, Inc.; p. 56, Dr. Ronald Shimek; p. 64, Marsh Youngbluth, Harbor Branch Oceanographic Institution, Inc.